I0462119

Scientific Proof that the Earth & Water Existed before Time, Space & the Big Bang!

Ifeanyi Chukwujama

This Booklet is available at:
www.amazon.com
www.kindle.com
www.createspace.com
www.truthnlife.com

For permission to reprint or copy this booklet, please contact the publisher:

Ifeanyi Chukwujama

Truth & Life Institute
Incorporated

394-396 Warren Street, Boston, MA 02119

Our email: JesusOnkm@gmail.com

Our website: www.truthnlife.com

ISBN: 10: 150050629X
ISBN-13: 978-1500506292

PRINTED IN THE USA

DEDICATION

To my children and my wife! I thank God Almighty for allowing me to live long enough to teach my children that the most important possession in this life is having an intimate relationship with God and His Son Jesus Christ through obedience and complete surrender of our lives to Jesus Christ!

Also to my readers who come to share the same value in Jesus Christ! May our merciful and loving God maintain His perpetual love in our hearts, and keep us under his glorious wings forever and ever. Amen!

CONTENTS

ACKNOWLEDGMENTS

The Bible says: *"Every word of God is flawless; he is a shield to those who take refuge in him."* *(Proverbs 30:5)*. I am a living testimony of that and I urge you to believe and experience the same thing for yourself.

The Bible also says that the gospel is not about talk but power. Anyone who believes in God and obeys His commands feels that power and would not allow anyone to talk them out of it with some new scientific facts.

From the same pages of the Bible I have read many times, something caught my eyes and very quickly the whole creation science began to unfold before my very eyes. It was almost as if I was injected into the middle of it all, and difference components of it were being explained to me. I had many questions, but I was promptly led to Bible passages to answer my questions.

The greatest step Christians could take to make disciples of the world would be to take Genesis seriously as the true historical records of life, the earth and the universe; and to teach it everywhere knowledge is taught.

The lie that is Evolution quickly got traction because its proponents were hard at work distributing books and holding press conferences about their fabrications. And within a few generation change-overs, their lies forced the true historical records in the Bible to the back burner.

And as they shoved their science fictions down everyone's threat, Christians sat in doubt or confused about the truth of creation, even though God took the pains to dictate the whole account to Moses, to give believers firm footing, since anyone who does not know his history would have a hard time charting his future.

Although most of Christianity would quickly tell you that Genesis is correct, only very few understand it, and fewer still have confidence in it. Christian education no longer places enough emphasis on the Genesis creation account as the irrefutable recorded history that it is.

Christianity has to renew its emphasis on Genesis because the whole book contains the science and history of creation and of human activities in the ancient times.

Genesis is not only scientific, it is super scientific, as it contains scientific maneuvers that are far beyond the understanding of the best scientific minds the world will ever produce.

Without the use of constants, the world of science would be at lost for words. Constants are used in science to substitute for information that no one has cooked up convincing explanation for, so that equations are solvable and lend credibility to scientific theories.

Gravity is the cohesive force exerted on the earth— and other like bodies throughout the universe—by the Spirit of God, to hold everything on the body down to its surface. But scientists, in their craze to exclude God from our world, explain gravity as coming from the earth and computed it at 9.8 m/sec^2. This constant works even though its nature is misinterpreted.

Scientific Proof that the Earth & Water Existed before Time, Space & the Big Bang!

Ifeanyi Chukwujama

Chapter 1

The Truth about Life and Nature

When we understand who God created the human being to be, we will then understand who God really is. This is how the human person is wired:

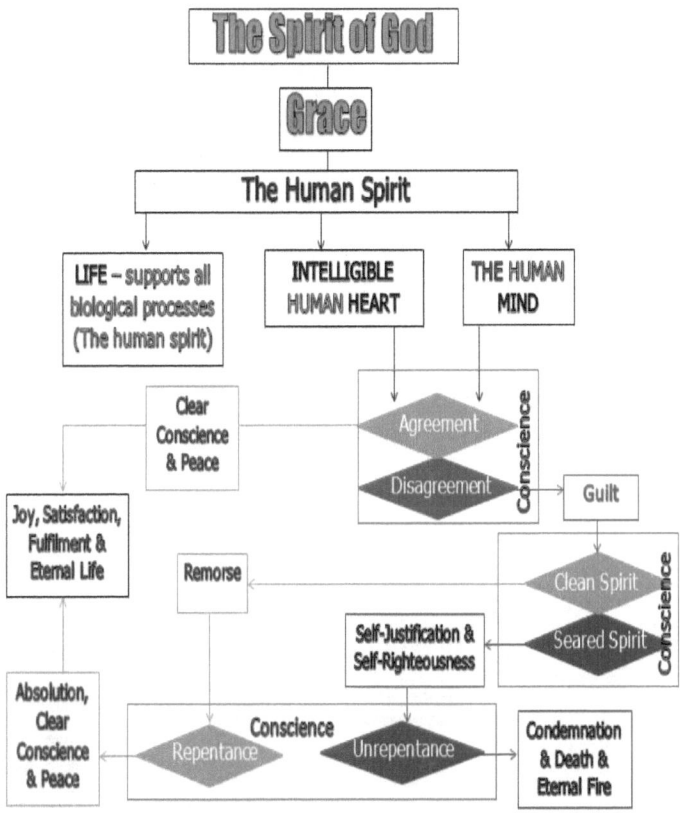

Figure 1: The human person has God written all over him, and filling him up completely. Everything the human person does is controlled by God in real time.

As well as this:

God's infinitely-advanced Cloud Technology Intelligence System

(John 10:30) God the Father
The Son & the Holy Spirit

Infinite LIFE

(John 1:1-5) & (Colossians 1:15-20)

GRACE

(God's Free Spirit—*(Psalm 51:12 KJV)*
(The God-Particle) *(Ephesians 4:3-10)*
- Moves at the speed of God, unimpeded
- The one particle that became everything

Constituent Grace

- Part of the constitution of all created natural & spiritual things (space, the earth, the universe, spirits, etc.) *(Ezekiel 37:4-8)*

Free Grace – Transient Grace

- **DARK ENERGY - 73% of the Universe**
- Passes freely through all of God's creations, at the speed of God
- Unimpeded, unabridged & universal
- The Breath of God *(Ezekiel 37:9-10)*

Spiritual Constituent Grace 23% of the Universe	**Naturalized Constituent Grace** 4% of the Universe	**Freely-Available Transient Grace**	**Saving Grace** Anointing Grace
- **DARK MATTER**	- **THE VISIBLE UNIVERSE**	- Completely saturates the universe & earth & all *Psalm 139:7-12 & Ephesians 4:10*	- Limited to **faith** in God & Christ *(John 1:11-13)*
- Angels & Demons	- everything	- repairs & renews	- The Grace that Protects, Guides, Repairs, **Exalts**, **Reconstructs** & Renews *(John 1:16)*
- Powers & rulers- *Ephesians 6:2*	- **Human Science Knows and could detect**	- moves at the Speed of God	
- **undetectable to Human Science**	and characterize	- The *juice of life*	
- moves at God-assigned speeds	- This constituent Grace is Nature *(Hebrews 11:3)*	- God's Treasury of Grace & *God's Seat of Mercy*	- Moves at the Speed of God
- God's platform of creation (space), & the frameworks of Grace	- moves at God-assigned speed	- God' **software** for Intelligence *Psalm 139:2-15, Job 32:8, John 14:10*	- **The substance of innocence**

The HARDWARE SYSTEMS   The Operating System & SOFTWARE

Figure 2: God controls both the life and the intellect of each and every human being, in real time.

God is infinite, and heaven is a literal place, sacred and majestic, which wholly encapsulates our universe. See below. Sitting comfortably on His throne in heaven, God scooped up the crystal clear water of heaven *(Revelation 22:1)* into the hollow of His mighty hand *(Isaiah 40:12)*, with the earth—a churning volcanic mass—fully submersed inside the water, spewing thick dark vapor in every direction (360°) into the space immediately surroundings the water in the hollow of God's mighty hand. Both the earth and the water were suspended in the hollow of God's mighty hand by the Spirit of God *(Job 26:7)*

Figure 3: Prior to Day 1, it was just water and the earth suspended

over God's mighty hand as pictured here *(Genesis 1:2)*. **On Day 2** *(Genesis 1:6-8)*, **Space was stretched out all around the suspended earth and water** *(Isaiah 45:12),* **with a deluge of water on top of space. See Figure 4 below. From Day 4 forward** *(Genesis 1:14-19),* **the suspended space/earth/water turned into the suspended universe** *(Isaiah 40:12)* **which is depicted in this diagram.**

The following diagram is an illustration of the suspended Space/water/earth at the end of Day 2 of God's creation in Genesis.

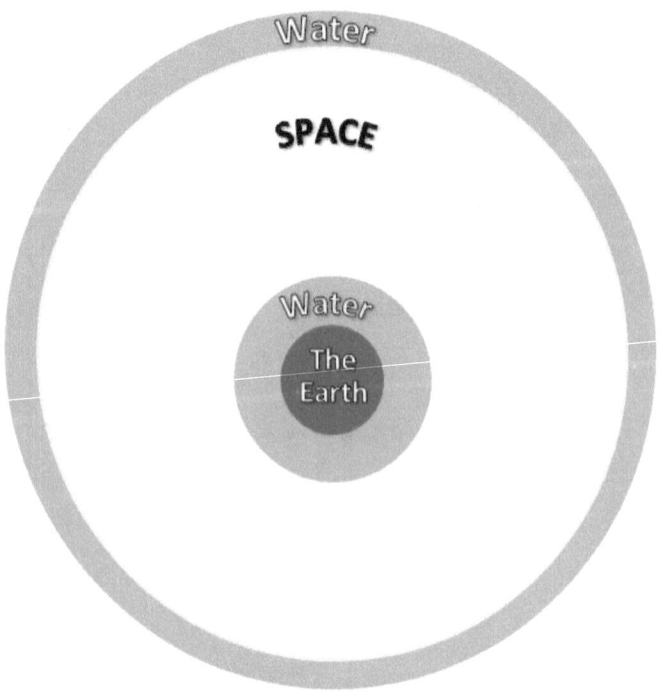

Figure 4: Earth, water and Space configuration at the end of Day 2 of creation.

The dense vapor created pitch darkness over the water, symbolizing the presence of Satan on the earth at the time of God's creation. Then God called for light on the submerged earth *(Genesis 1:3-5),* and light pierced through the thick darkness which was covering the earth and the water in the hollow of God's hand, kicking off time as we know time today.

The very first appearance of light on the surface of the earth and the water in the hollow of God's hand is the dawn of time—**high noon.** Evening followed; and then, the morning; to close out the very first day ever, on the planet earth!

Time started at high noon, and God started counting time at the very moment of light's first appearance on the earth. Science does not get any more precise that that! God's counting of time from the very instant of light's first impingement on the surface of the earth is exactly the reason why the Bible concludes each day of creation in Genesis Chapter One, with the authoritative declaration:

"And there was evening, and there was morning ..." *(Genesis 1:5); (Genesis 1:8); (Genesis 1:13); (Genesis 1:19); (Genesis 1:23) & (Genesis 1:31).*

The world, however, started counting time from midnight (12:01 a.m.), when it is pitch darkness. Time on the earth did not start in the midst of darkness. The Bible did not state how long darkness covered the earth and the water in the hollow of God's hand before God called for light to appear on the earth. Mankind has no business starting time at midnight—in pitch darkness. It was light that ushered in time on the planet earth; and not darkness!

Who calls time correctly—God, who started time, and maintained record of time with unsurpassed precision; or the world and its scientists, who arbitrarily chose midnight to begin the day? You be the judge!

Yet, the world and its scientists and the atheists call the Bible claims baseless; and successfully convinced humanity to walk away from the Bible and all of the Bible's truths. All mankind must return to trusting their individual spirits, as opposed to trying to reinvent what God had already clearly and unambiguously posited for mankind on the pages of the Bible. There is really nothing more to add to what God has already documented!

The light, which appeared on the earth at God's command in *Genesis 1:3,* and all of its associated radiant energies, is none other than Jesus Christ the Son of God. This is why the gospel says:

"In the beginning was the Word, and the Word was with God, and the Word was God. ² He was with God in the beginning. ³ Through him all things were made; without him nothing was made that has been made. ⁴ In him was life, <u>and that life was the light of all mankind.</u> ⁵ <u>The light shines in the darkness, and the darkness has not overcome it.</u>" (John 1:1-5)

"Out of his fullness we have all received grace in place of grace already given. ¹⁷ For the law was given through Moses; grace and truth came through Jesus Christ. ¹⁸ No one has ever seen God, but <u>the one and only Son, who is</u>

__himself God__ and is in closest relationship with the Father, has made him known." (John 1:16-18)

Since life is Grace, the proclamation, *"Out of his fullness we have all received grace in place of grace already given,"* paraphrased, is saying: that out of the fullness of Jesus Christ, believers received life in place of the life already given to each human being at conception by God inside the womb of their mother.

Not only is Jesus Christ the giver of the abundant life *(John 10:10)*, He is also the giver of the original life, which every human being received in their mother's wombs *(John 15:5)*. That is why the Bible declares that Jesus Christ holds everything together, and that everything was made in Him, through Him and for Him *(John 1:3)* & *(Colossians 1:15-20);* and that in Him, all human beings move and act and have their being *(Acts 17:28)*. Every human body is a temple of God because the Spirit of God dwells inside each person in the form of human spirit (Grace).

The entire account of creation in the Book of Genesis is precision in itself—nothing more, nothing less. The earth and the water were created by God before anything else in nature was created. That is exactly why the Genesis creation account does not cover the creation of either water or the earth. Both were starting materials for God's creation activities of Genesis.

Light also was in existence prior to God's creation

11

account in Genesis. God's creation narrative in Genesis started with sizzling hot, formless earth, submersed inside the water of creation in the hollow of God's mighty hand—both matter in pitch darkness. God then called for light to dissipate the darkness, and usher in time on the planet earth: On Day 1, God called light to shine on the earth and on the water; and time began.

Then, on Day 2, God created space, and all of its properties, including orbital configurations, orbital speeds and orbital orientations. On Day 4, God populated space, to create the universe; obscuring earth's view of God and God's majestic heaven: Mankind's sinfulness would cause any human being that sees God to perish instantly *(Exodus 33:20)* & *(Leviticus 16:2)*.

But we have all been taught that the universe came of its own, due to gravity—when gravity, in actuality, is the Spirit of God, who the Bible proclaimed was moving *"upon the face of the water of creation"* at *Genesis 1:2 KJV.* The Spirit of God accomplished every command of God in Genesis; bringing forth each creation as soon as the word left the mouth of God. And the Spirit of God has continued to hold all of God's creations together from the very moment of their respective creations up until this moment. This is why the Bible says that in Jesus Christ all thing are held together *(Colossians 1:17).*

In essence, the Spirit of God created everything through God's spoken word—the Word—who later became flesh *(John 1:14),* and came into the world and dwelt among human beings give them redemption from sin. The world scientists' claim that gravity created the universe is technically accurate; except that they are ignorant of what gravity truly is. Such an ignorance means singing the praises of something which you are at the same time rejecting and beating up on!

On the first day of creation, God brought forth the **light,** and darkness receded. Light was not created on this day. Light was made to shine on the earth for the first time on this day. And God was the source of that light—*"God is light; in him there is no darkness at all" (1 John 1:5);* which makes absolute sense; because the Origin of light Himself (God) and the heavenly hosts who were present at the creation of the earth and the universe would not have been in darkness before this time.

Therefore God has had light for all eternity because that is His nature. And God promises to be mankind's direct light again at the end of time; when the current universe passes away and the earth is renewed, *Revelation 22:5.*

God shined on the earth from Day 1 to Day 3, through His Son Jesus Christ *(John 1:1-5).* God's direct radiance, which illuminated the earth from Day 1 to day 3 of creation, was precision-targeted on the earth for God's creation accounting purposes; because God's full radiance would have instantly flushed out even the thickest darkness from all 360° (360 degrees) of the earth—given the earth's insignificant size in comparison to the infinite size of God who was the source of that light.

God's Dwelling Place

(HEAVEN)

God's Eternal Luminance fills the whole heaven 24/7

This Diagram is Not Drawn to Scale!

This is the Universe, suspended by God's Free Spirit (Gravity) over God's Mighty Hand, and Rotating at a speed that is faster than the speed of light

Figure 5: All of heaven is flush with God's radiance, eliminating any hints of darkness anywhere in heaven. That is why heaven is timeless!

To verify this for yourself, look at a tennis ball on a bright sunny day, and notice that all 360° of the tennis ball is lit by the sunlight because of the tennis ball's small size as compared to the huge sun which is casting the light. Therefore, only a precision targeted light beamed onto the earth from an infinite source of light such as God, would leave one half of the earth in darkness to form the nighttime.

And only a precision targeted light beamed on the earth, that creates equal length of daytime and nighttime would create high noon, evening and morning, as recounted in God's creation narrative of

Scientific Proof that the Earth & Water Existed before Time, Space
& the Big Bang!

Genesis Chapter One—as illustrated in the diagram
below.

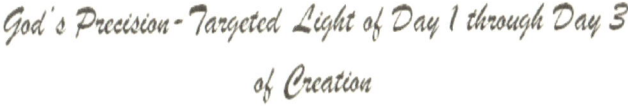

God's Precision-Targeted Light of Day 1 through Day 3

of Creation

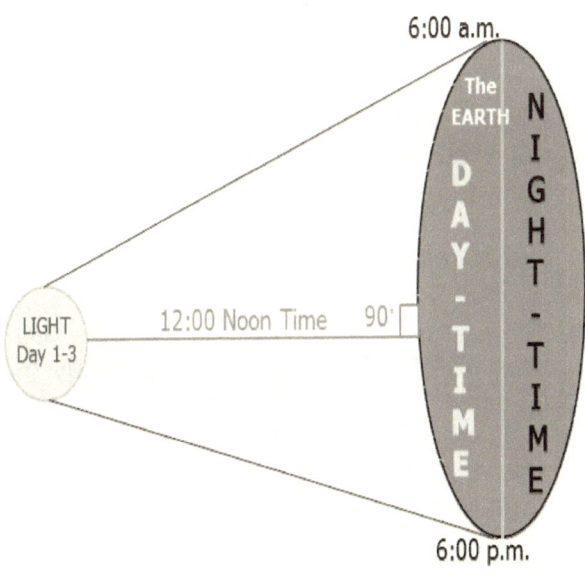

**Figure 6: God configured the light that shone on the earth on Day 1 of
creation to illuminate only half the earth and leave the other half in
darkness. God's full luminance leaves no darkness anywhere as
illustrated in this diagram. The meridian on the earth that is
perpendicular to the surface of the earth, at light's first appearance on
the earth on Day 1 of creation experience high noon at the instant of
lights appearance on the earth; with 6 hours of morning time to one
side of the perpendicular meridian; and six hours of evening time on
the other side of that perpendicular meridian.**

The precision-targeted direct radiance of God on Day 1 of creation, lit up only one half of the sphere that is the earth, and left the other half in darkness, causing God to announce proudly that the point of light's first impingement of the earth, which is perpendicular to the focus of the light on impact, reflects high noon—with 6 hours of morning to one side and 6 hours of evening to the other.

The earth started to rotate so that every portion of the earth would experience high noon within a 24-hour period. The first point of high-noon on the earth at the light's appearance then experienced **evening** time, followed by **morning** time; and then back to high noon, to complete the first day.

The earth's gases started to form through ionization by the True Light. The Bible says that God is a consuming fire *(Hebrews 12:29—"for our God is a consuming fire")*. Therefore no celestial body has the ionizing power that equals God's—God is infinite energy plus the sum total of all the energies that exist everywhere in the universe, known and unknown, visible and invisible.

The dawn of light on the earth marked the beginning of **time** in the universe since the earth was created before time, and the universe was created on Day 4 of creation.

The Bible infers that as the celestial bodies raced away in every direction from around the earth at their creation on Day 4 of creation, God arranged them in clusters to form galaxies; set the galaxies and the individual bodies within each galaxy in orbits, and their inter-relationships; and set their motions and functions.

While God was creating the universe through extreme energies, He protected the fragile earth and its completed infrastructure from being vaporized by the excessive energies generated in the Big Bang; thereby putting in display the limitlessness of God's capacities—He is God and there is no other!

Galileo is widely hailed for postulating that the earth and the other planets are revolving around the sun, thereby making the sun the center of our solar system. And the world quickly dismisses those who believed that the earth is the center of the universe. The sun only became the center of our solar system after the sun was created by God on Day 4 of creation.

The truth remains that the earth is the center of the universe. In the beginning, the earth—and the water surrounding it—was the only thing is orbit because there was no universe at that point. The earth started to rotate on the instant of light's first appearance on the surface of the earth—*Genesis 1:3-5*—*"Let there be light," and there was light … <u>And there was evening, and there was morning—the first day."</u>*

The meridian perpendicular to God's focus of the light at light's first appearance on the earth immediately experienced high noon. Then as the earth rotated for the next twenty four hours, *"there was evening, and there was morning,"* and finally, a second high noon that kicked off Day 2 of creation.

The earth remained the only body in orbit for three full days before the sun, the moon, the stars, and the universe, were created through the Big Bang on Day 4 of creation. After the sun was created, the earth began to revolve around the sun.

And it is tradition with God throughout the Bible to carry over everything He started without losing one beat. Secondly, our solar system being extremely small in size when compared to the whole universe still makes the earth the center of the universe. With God, it's all about common threads.

God's common thread on the animal world got Charles Darwin and all the evolutionists all twisted and out of sync, causing them to extrapolate minor variations into non-realities; with which they confused the entire human society—thereby denying people of the grace that was given freely to all, causing them to hurtle towards the eternal lake of fire.

Here is God's confidence about His creation activities. As God was crafting His masterpiece, the angels watched and celebrated: Here is the Scripture:

"Where were you when I laid the earth's foundation? Tell me, if you understand. Who marked off its dimensions? Surely you know! Who stretched a measuring line across it? On what were its footings set, or who laid its cornerstone—while the morning stars sang together and all the angels shouted for joy? (Job 38:4-7).

We learned from the Bible that God is light and there is no darkness in Him whatsoever *(1 John 1:5)*. Why then, did God start the earth filled with darkness? Everything God does have meaning beyond human comprehension. Therefore, for God, who has no darkness in Him whatsoever to start the world with darkness speaks volumes. That is the kind of stuff God had designed for mankind to pursue.

There was infinite space between the water in the hollow of God's hand and God's face. That is more than enough space to dissipate the vapor that rose from the quenching of the much smaller molten mass of earth in the water in the hollow of God's hand.

But God chose to concentrate the vapor that rose from the quenching of earth right around the water in the hollow of His hand, until it became pitch black all around the water, as reported in *Genesis 1:2*.

The darkness that covered the water and the earth in *Genesis 1:2* was symbolic of the darkness of the heart that would soon overtake mankind, because God created the earth for the benefit of mankind. That darkness of the heart would come from Lucifer, whose burnt out remains became reconstituted by God to form the earth—*Ezekiel 28:8*—*"They will bring you down to the pit, and you will die a violent death in the heart of the seas."*

Here is the rest of the passage about Lucifer's destruction; and the birth of the earth:

"You were the seal of perfection,
  full of wisdom and perfect in beauty.
[13] You were in Eden,
  the garden of God;
every precious stone adorned you:
  carnelian, chrysolite and emerald,
  topaz, onyx and jasper,
  lapis lazuli, turquoise and beryl.
Your settings and mountings were made of gold;
  on the day you were created they were prepared.
[14] You were anointed as a guardian cherub,
  for so I ordained you.
You were on the holy mount of God;
  you walked among the fiery stones.
[15] You were blameless in your ways
  from the day you were created
  till wickedness was found in you.
[16] Through your widespread trade
  you were filled with violence,
  and you sinned.
So I drove you in disgrace from the mount of God,
  and I expelled you, guardian cherub,
  from among the fiery stones.
[17] Your heart became proud
  on account of your beauty,
and you corrupted your wisdom
  because of your splendor.
So I threw you to the earth;
  I made a spectacle of you before kings.
[18] By your many sins and dishonest trade
  you have desecrated your sanctuaries.
So I made a fire come out from you,
  and it consumed you, and I reduced you to ashes on the
ground in the sight of all who were watching." (Ezekiel
28:12-18)

Lucifer was a morning star—the brightest burning star, adorned with all the precious stones that there was. But pride ruined it all for him. God made a more intense fire come out from within the fiery star that was Lucifer, and the fire burnt Lucifer to ashes, which was quenched by the water in the hollow of God's hand; leaving the core of Lucifer's burnt out remains still churning furiously.

An intense fire dropping into a deluge of water is sure a violent death for the fire. The quenching of such fire would be intense sizzle and boiling over of water and a blanket of thick dark vapor that creates darkness on the surface of the water, as reported in *Genesis 1:2.*

A fire that falls into a sea of water is destined for a sudden death. And that was exactly how Lucifer was extinguished from the shining star to a burnt out star that became a planet. For further humiliation, God reconstituted Lucifer's burnt out remains and made it the earth—a habitation for God's natural creations, who would be stepping all over the remains of Lucifer—in in ultimate triumph over wickedness.

For final details as to how an eternal soul that was Lucifer was destroyed because of wickedness, God added the following:

"Have you ever given orders to the morning,
or shown the dawn its place,
<u>13 *that it might take the earth by the edges*</u>

> *and shake the wicked out of it?*
> [14] *The earth takes shape like clay under a seal;*
>   *its features stand out like those of a garment.*
> [15] *The wicked are denied their light,*
>   *and their upraised arm is broken.*" *(Job 38:12-15)*

The preceding passage says that God shook Satan out of the earth; and made the earth habitable— *Isaiah 45:18*—"*For this is what the LORD says— he who created the heavens, he is God; he who fashioned and made the earth, he founded it; he did not create it to be empty, but formed it to be inhabited— he says: "I am the LORD, and there is no other."*

God's shaking of Satan out of the earth to create a safe habitation for mankind, and all of God's natural creations, is what God was referring to in *Genesis 2:17*—"*but you must not eat from the tree of the knowledge of good and evil, for when you eat from it you will certainly die.*"

With this sanction, God forbade mankind from seeking counsel from Satan, who was at the Garden of Eden at the time—*Genesis 2:9*—"*In the middle of the garden were the tree of life and the tree of the knowledge of good and evil.*"

But if of a person's accord, they choose to deal with Satan, the person's decision automatically unbinds Satan. And Satan could do to the person whatever Satan chooses. God was also alluding to the two components of the human intellect in *Genesis 2:9*—the human spirit (*"heart"*) and the human mind.

At the point in time of *Genesis 1:2*, there was heaven (God's dwelling place), with God majestically seated on His throne, and the earth sizzling inside the water in the hollow of God's hand: *"The devil was hurled to the earth"* *(Revelation 12:9)*—the voided persona of Lucifer—after Lucifer was stripped of Grace and kicked out of heaven.

God had reserved the red-hot core of the earth as the devil's final destination—God had reduced the once mighty Lucifer to a hollowed being who would not spend eternity in a cramped up earth's core which was only a small part of his previously mighty self. Therefore, the devil—a hollowed spirit—was on the earth, a step closer to his ultimate fiery abode. See the following passage from the Bible:

"How you have fallen from heaven, morning star, son of the dawn! You have been cast down to the earth, you who once laid low the nations!" (Isaiah 14:12)

"For thou hast said in thine heart, I will ascend into heaven, I will exalt my throne above the stars of God: I will sit also upon the mount of the congregation, in the sides of the north." (Isaiah 14:13, KJV)

"I will ascend above the tops of the clouds; I will make myself like the Most High." ¹⁵ But you are brought down to the realm of the dead, to the depths of the pit." (Isaiah 14:14-15)

"Those who see you stare at you, they ponder your fate: "Is this the man who shook the earth and made kingdoms tremble, ¹⁷ the man who made the world a wilderness, who overthrew its cities and would not let his captives go

home?"" *(Isaiah 14:16-17)*

Before God cast the Lucifer out of heaven as Satan, God had already decided to reconstitute Lucifer's burnt-out remains as the earth. So when God cast Lucifer out of heaven, He cast the burning Lucifer into the water in the hollow of God's mighty hand, and jicked off creation narrative of Genesis One.

The cooling by the water of the churning mass inside the water of creation produced a solid crust all around the submersed earth, with the core of the earth still red-hot and churning violently. God designed this fiery core of the earth as the devil's final destination through all eternity. The solidified crust on the surface of the submersed earth became the earth's sea beds.

At this point in creation, there was our infinite God on His holy and majestic throne in heaven, with the heavenly hosts attending to Him 24-seven. And the heavenly hosts spread out across the landscape of heaven. There was water in the hollow of God's mighty hand. And there was this blob of shapeless mass (the earth) submersed inside that water in the hollow of God's hand.

And this mass was churning mad because it was an eternal volcano, created as the devil's eternal home, because the devil had entertained wicked thoughts against God Almighty *(Isaiah 14:13-14)*. God cooled off the crust of this volcanic blob, and later, caused land to project from the blob above the surface of the water, so that God could put the humans on the land; reserving the blazing core of it for the devil and all the enemies of God.

The thick dark vapor from the quenching of the earth's submersed surface threw the surface of the water and the earth into thick darkness. This was the starting point *(Genesis 1:2)* in God's narrative about His creation activities.

And because the earth's surface, and the surface of the water surrounding the earth, is in pitch darkness, God called for light to pierce through that darkness and light up the earth and the water *(Genesis 1:3)*.

And light cut through the darkness, which promptly receded to one half of the sphere that is the earth; and its water covering. That light came directly from God Himself—the great God who is all light *(1 John 1:5)*.

This is exactly what *John 1:1-5* was talking about. That light that cut through the darkness over the surface of the earth and the water was Jesus Christ the Son of God, who was there in the beginning with God. Through Him and for Him, everything that was created was created. Jesus Christ came onto the earth at *Genesis 1:3* and the devil was pushed out of one half of the earth's surface; and continued to be chased around by the light—day after day.

Therefore, from that first impingement of light on the surface of the water, and of the earth, the world had been split between light and darkness—and between good and evil.

Yet, when God created mankind on Day 6 of creation, God was sure to make all evil inert to mankind so that mankind would be free from sinful nature.

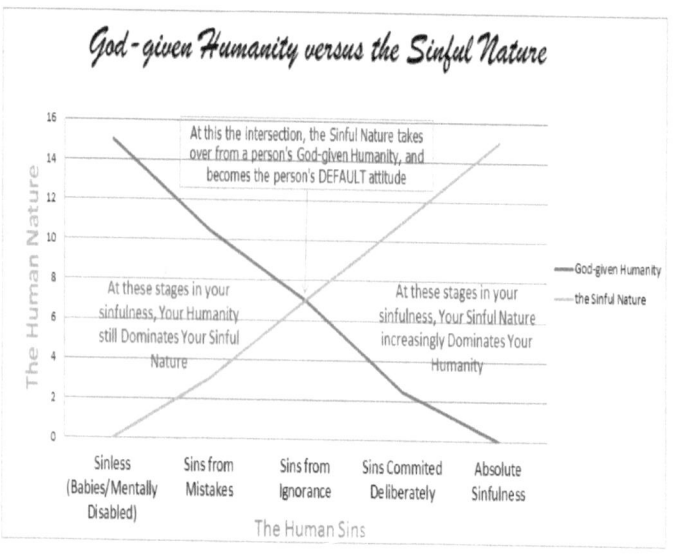

Figure 7: Grace covers over all evil when we work to maintain the Grace which God had endowed us with.

God gave mankind one simple command to get mankind to maintain the discipline required to justify the freedom mankind was given as dictated by love. Here is that single command of God to mankind:

"You are free to eat from any tree in the garden; [17] but you must not eat from the tree of the knowledge of good and evil, for when you eat from it you will certainly die." (Genesis 2:16-17)

The appearance of light on the earth caused the

earth to rotate so that every part of the earth's surface would see the light within every twenty four hour period.

The Bible also told us that God spreads out the northern sky over empty space *(Job 26:7)*. Since the earth is spherical in shape, everywhere you are on the earth has a northern sky. This passage then translates to God spreading out the sky uniformly across the face of the sphere that is our earth.

And *Genesis 1:6-8* told us that God created our space and separated waters from waters. Here is that passage:

"And God said, Let there be a firmament in the midst of the waters, and let it divide the waters from the waters. [7] And God made the firmament, and divided the waters which were under the firmament from the waters which were above the firmament: and it was so. [8] And God called the firmament Heaven. And the evening and the morning were the second day." (Genesis 1:6-8 KJV)

God introduced space all around the sphere that is our earth, and uniformly pushed the space away from the earth, to put distance between the surface of the earth and the water remaining over the face of the earth, and the water God was pushing away from the face of the earth, which was atop the newly created space. This water God pushed out of space remains outside space today. That is the crystal clear water of Heaven which the Bible mentioned a few times.

Ionization of atmospheric gases started after the creation of space on Day 2. And the earth's atmosphere was stratified by God, perfected and sealed off from the rest of space before the Big Bang occurred on Day 4 of creation.

It can be observed from *Genesis 1:6-8* that God spent the entire second day of creation on the development of space and all of its properties, including orbital configurations, orbital speeds and orbital orientations.

On this second day of creation, God mapped out space, and made allocations for stellar and planetary placements and distributions, stellar configurations and motions and interrelations. Slots for all the various galaxies were designated, set and perfected, in preparation for the Big Bang that was to come on Day 4 of creation.

On Day 3 of creation, at God's command, an unprecedented volcanic eruption parted the still waters that held the earth, creating a huge landmass out of the water, triggering huge tsunamis that clapped violently over the new landmass, cooling the landmass and weathering it.

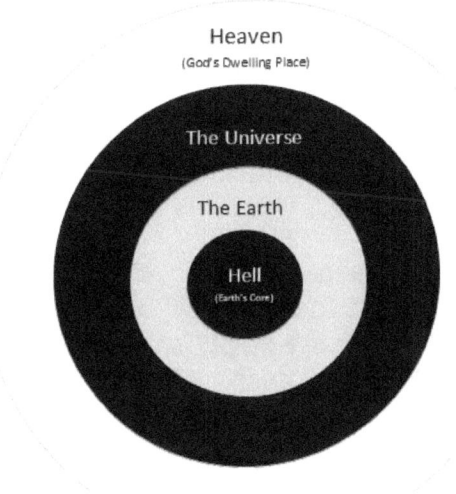

Figure 8: These are the Circles of Life (Not drawn to proportion) - This is the final configuration—of heaven, the universe, the earth and hell—after God created everything on Day 6 of creation. First there was God sitting majestically on His holy throne in heaven, with the earth submersed in the water in the hollow of God's mighty hand. Then God started His creations as detailed in Genesis Chapter One, by first making light appear on the surface of the earth to push the darkness on the surface to one side, and chase the darkness around throughout each day. Then God created our expansive space over the earth, and pushed water out of space as He stretched out space, and left some water on the surface of the earth to become earth's seas. Then God created land out of the earth's seas, cooled and weathered the land and planted green vegetation of the land. Then God caused a cataclysmic explosion that created our sun, moon and stars, and every other celestial body in the universe; and populated space with all of them. Then God created sea creatures and flying creatures. And God created land creatures and finally mankind. Then God rested.

Huge reservoirs of water were trapped inside the new landmass as the volcano jetted through the water and cooled. These reservoirs opened up into rivers, lakes and ponds later that day to provide potable water for mankind and the animals. And God put green vegetation on the land—supported by the light that dawned on the earth on Day 1 of creation; the ionized gases that formed the earth's atmosphere; water from the rivers and lakes; and nutrition from the new earth's crust.

At God's command on Day 4 of creation, a cataclysmic explosion rocked the earth and reverberated across the empty space that surrounds the earth, creating oceans of fires and radiant energies that raced across space at the speed of God, and dispersed throughout space to form the universe—with the Spirit of God completely shielding the fragile earth's ecosystem and atmosphere from the unprecedented explosion.

It should be noted here that God did not leave His throne to accomplish any of these. All of the creation events and activities were accomplished by God, on the palm of God, while God was comfortably seated on His throne in heaven:

"The Lord is in his holy temple; the Lord is on his heavenly throne. He observes everyone on earth; his eyes examine them." (Psalm 11:4)

As expansive and as awe-inspiring as space—and our universe—are, that entire expanse fits into the palm of God; and is suspended over God's palm by the Spirit of God. God sat at His throne and accomplished all He wanted to accomplish. And it was all perfectly done.

Day 4 of creation was the beginning of a new age on the earth. While Day 1 through Day 3 had God directly supplying His light to the earth, on Day 4 the sun replace God's direct light on the earth. And this will continue till the end of time.

It is human ignorance that made the world scientists think that they could figure out how the earth and the universe started, without full acknowledgement of God. Nobody who disrespects God ever gains glory from God. God declared in the Bible that human wisdom is foolishness to God. The truth of everything can only be learned through humility and obedience to God. God leads anyone who antagonizes God further away from the truth—and not to the truth:

"The wrath of God is being revealed from heaven against all the godlessness and wickedness of people, who suppress the truth by their wickedness, [19] since what may be known about God is plain to them, because God has made it plain to them. [20] For since the creation of the world God's invisible qualities—his eternal power and divine nature—have been clearly seen, being understood from what has been made, so that people are without excuse.

[21] For although they knew God, they neither glorified him as God nor gave thanks to him, but their thinking became futile and their foolish hearts were darkened. [22] Although they claimed to be wise, they became fools [23] and

31

exchanged the glory of the immortal God for images made to look like a mortal human being and birds and animals and reptiles.

[24] Therefore God gave them over in the sinful desires of their hearts to sexual impurity for the degrading of their bodies with one another.[25] They exchanged the truth about God for a lie, and worshiped and served created things rather than the Creator—who is forever praised. Amen.

[26] Because of this, God gave them over to shameful lusts. Even their women exchanged natural sexual relations for unnatural ones. [27] In the same way the men also abandoned natural relations with women and were inflamed with lust for one another. Men committed shameful acts with other men, and received in themselves the due penalty for their error.

[28] Furthermore, just as they did not think it worthwhile to retain the knowledge of God, so God gave them over to a depraved mind, so that they do what ought not to be done. [29] They have become filled with every kind of wickedness, evil, greed and depravity. They are full of envy, murder, strife, deceit and malice. They are gossips, [30] slanderers, God-haters, insolent, arrogant and boastful; they invent ways of doing evil; they disobey their parents; [31] they have no understanding, no fidelity, no love, no mercy. [32] Although they know God's righteous decree that those who do such things deserve death, they not only continue to do these very things but also approve of those who practice them." (Romans 1:18-32)

God's questioning of Job clearly shows that unless a human person has faith in God, and trust in God's word, the person can never get to the truth of anything God did way before the person was born. Everything the world scientists have concocted and fed to the world about the origin of the earth and the universe is mere science fiction. That they have been able to take pictures of celestial worlds and distribute them to the public, lends credibility to their falsifications. But they have been looking at the universe backwards.

Here is God rebuking Job about projecting that he knew what he did not know:

"What is the way to the abode of light? And where does darkness reside? *20 Can you take them to their places? Do you know the paths to their dwellings?* *21* **Surely you know, for you were already born! You have lived so many years!"** *(Job 38:19-21)*

What the underlined portion of the preceding passage is saying is that unless you personally witnessed something, whatever explanation you come up with could be all lies. If you did not personally witness something and gain correct knowledge of what really took place, you must base your conclusion on a reliable source to have a chance at the truth.

And what source anywhere in the universe is more reliable than the very words of the Creator, who put these things together in the first place? Mankind's great desire to know figure everything out on its own, and not take God's word for it, continues to be mankind's greatest liability.

Instead of allowing our minds to be enlightened by knowledge, we continue to let knowledge ruin our lives. Any knowledge that ruins anyone's chance of salvation is not knowledge but a seed of the devil. We all have the "Eve Syndrome": why take God's word for it? But doubt corrodes the body and the mind!

The devil has been deceiving human beings since Adam and Eve. He always finds a way to tell mankind something that is contrary to what God has told mankind. We look at it as brilliance, but it is a path that leads to death and destruction. My seventeen year old son told me that when the devil tells us things, he makes us believe that we came up with the things on our own, so we could pat ourselves on the back and add to our feeling of self-importance. The devil is not interested in taking credit for a wrong that any human person does, because he knows the reward for wrongs. He is interested in getting more people on his side, to prove to God that he is not the only one who does evil:

""Does Job fear God for nothing?" Satan replied. [10] "Have you not put a hedge around him and his household and everything he has? You have blessed the work of his hands, so that his flocks and herds are spread throughout the land. [11] But now stretch out your hand and strike everything he has, and he will surely curse you to your face." (Job 1:9)

If the devil lets you know that it is him who is filling your mind with thoughts and ideas that contradict the word of God, and that he is doing it just so that you can confront God and prove the devil right; you will tell the devil outright, to go fight his own battle by himself, and leave you out of it.

That is why the devil fills your head with rebellion against God, and allows you to take credit for coming up with them as bright and exciting ideas, so that you can go ahead and execute them to your peril. That is why the devil is the master of disguise and deceit *(2 Corinthians 11:14-15)*.

Every account of creation out there that is contradictory to the very simple and straightforward account of creation in the Book of Genesis is an innuendo and a flat out lie. The so-called origin scientists are basing their theories on assumptions, endless extrapolations and billions of years of formation. But in truth, the world had been through different stages of existence, over a relatively short period of time, because the world's architect and builder is the master of perfection:

"My own hand laid the foundations of the earth, and my right hand spread out the heavens; when I summon them, they all stand up together." (Isaiah 48:13)

The first three days of creation is very different from Day 4 of creation onwards. Day 1 through Day 3 of creation constituted the **1st age of the earth**. From Day 1 to Day 3 of creation, the only thing standing between the earth and God is the water in the hollow of God's hand. And that water is crystal clear and would not have obstructed anyone's view.

But God chose to allow the vapor that rose from the

quenching of the volcanic mass that was the earth inside the water in the hollow of God's hand, to concentrate into a thick dark cloud that blanketed the entire surface of the water, and cut off God's light from the earth and the water. This was symbolic of the presence of the devil on the earth at that point.

God then called for light on the earth and God's direct light cut through the darkness and pushed the darkness to one half of the earth. The appearance of God's direct light on the earth caused the earth to rotate so that every portion of the planet would see God's direct light within each twenty four hour period.

The very instance of the first impingement of God's direct light on the face of the earth was high noon. Evening rolled in, followed by nighttime. And the morning closes out the day and leads to the next high noon. That is why the Bible concluded each day of creation with the following statement:

"And there was evening, and there was morning …"

Everything God does, the world does in reverse. God who created everything started counting time from high noon—beginning every day at noon and ending every day at noon. But the world changed the day to start at midnight and end at midnight, because the world believes that it knows better than God.

The world scientists claim that their knowledge is more precise than the information in the Bible. Is it really? Who is scientifically accurate in the counting of time here—the scientist's or the God of creation? Light showed up on the earth for the first time at high noon and God starting counting time from that very moment of first impingement of light on the surface of the earth—perpendicular to the surface of the earth at the point of that first impingement. But our current counting of time places the start of the day at a time of pitch darkness--midnight. And we call that scientific precision?

The first appearance of light on the earth, high noon——not the middle of the night—ushered in the first day ever on the earth. And the Bible calls it correctly—to the nearest nanosecond.

In this comparison we can all see that the Bible's account and record of light's first appearance on the earth is accurate to the dot, while the world's was simply a fabrication chosen for convenience and political purposes.

The problem has not only been the world. The greater problem has been the believers of God who have refused to use the knowledge God has made available to the world to develop better understanding of the precise truths God had posited for His people on the pages of the Bible. When you glance through the passages of the Bible and overlook phrases like *"And there was evening, and there was morning,"* you will fail to pick up the science of God's truths in the Bible.

God's direct light graced the face of the earth for the first three days of creation in Genesis Chapter One.

This was the maiden age of the earth—the perfect earth. And this age of the earth would return to the earth at the end of time as prophesied in the Book of Revelations *(Revelation 21:1-27)* & *(Revelation 22:1-5)*. Here are portions of that passage:

"I did not see a temple in the city, because the Lord God Almighty and the Lamb are its temple. ²³ **The city does not need the sun or the moon to shine on it, for the glory of God gives it light, and the Lamb is its lamp.** ²⁴ *The nations will walk by its light, and the kings of the earth will bring their splendor into it.* ²⁵ *On no day will its gates ever be shut, for there will be no night there.* ²⁶ *The glory and honor of the nations will be brought into it.* ²⁷ *Nothing impure will ever enter it, nor will anyone who does what is shameful or deceitful, but only those whose names are written in the Lamb's book of life." (Revelation 21:22-27)*

"Then the angel showed me the river of the water of life, as clear as crystal, flowing from the throne of God and of the Lamb ² *down the middle of the great street of the city. On each side of the river stood the tree of life, bearing twelve crops of fruit, yielding its fruit every month. And the leaves of the tree are for the healing of the nations.* ³ **No longer will there be any curse.** *The throne of God and of the Lamb will be in the city, and his servants will serve him.* ⁴ *They will see his face, and his name will be on their foreheads.* ⁵ **There will be no more night.** **They will not need the light of a lamp or the light of the sun, for the Lord God will give them light.** *And they will reign for ever and ever." (Revelation 22:1-5)*

At the end of time, there will be no more nights because there will be no more curses on the earth: The devil is gone forever, and would no longer torment humanity. And if there are no nights, time will become an endless continuum that stretches infinitely. This is the eternity the Bible promises believers.

When on Day 4 of creation, God replaced His direct light with the sun, the moon and the stars, God continued to shield the earth and all of its content by His presence through Grace which constitutes the bulk of God's creation. The devil was on the earth but the devil could not affect anything on the earth, because the devil and his minions were completely contained by God, and tucked away from God's perfect creations.

God continued to give His light to the world via the sun, the moon and the stars. Nevertheless, on Day 4 of creation, the heat of the sun was not scorching, like it does in our current world. And there was no rain or any inclement weather conditions to thwart the comfortable living conditions that existed in the world during this period.

The world was still perfect, even with the sun replacing the direct light of God. God had perfectly transitioned the earth's reliance on His direct luminance to the radiant energies of the sun, the moon and the stars; without the earth or anything on the earth, losing a beat.

This is why the world had missed this information. Most believers, and theologians, simply gloss over this transition without realizing its significance; thereby missing its impact on science and on our relationship to God and His Christ.

Revelation 7:16 is a window through which we can see the conditions on the perfect earth which God had originally created for humanity's perfect habitation. In the following passage, the people of the world who lived a life of faith and obedience to God and His Christ had been washed of their sins and renewed in holiness so they could live in everlasting blissfulness with God and His Christ.

The presence of God will shelter them from all adversities, including the scorching heat from the sun and other adverse weather conditions. That is a window for the world into the holy lives Adam and Eve lived with God at the Garden of Eden before they rebelled against God. Mankind could not derive a true understanding of the life conditions of the holy Adam and holy Eve any other way. Only God gives any human being true access to any event that ancient. Here is that window:

"These are they who have come out of the great tribulation; they have washed their robes and made them white in the blood of the Lamb. [15] Therefore, "they are before the throne of God and serve him day and night in his temple; and <u>he who sits on the throne will shelter them with his presence. [16] 'Never again will they hunger; never again will they thirst.</u> **The sun will not beat down on them,' nor any scorching heat.** *[17] <u>For the Lamb at the center of the throne will be their shepherd; 'he will lead them to springs of living water.' 'And God will wipe away every tear from their eyes.'"'"* (Revelation

7:15-17)

The 1st age of the earth—Days 1-3 of creation—was characterized by the direct presence of God on the earth. It will actually be accurate to call this age of the earth, Christs' first coming to the earth. Jesus Christ was the light that dawned on the earth on Day 1 of creation. The Bible confirms this claim in the following passages: *John 1:1-5, Colossians 1:15-20, Acts 17-28, Ephesians 4:4-6, etc.*

Age of the Earth	Period of Time	State of God's Creations on the Earth	Notable Events or Conditions on the Earth
1st	From Day 1 of Creation to Day 3 of Creation	Perfection	• God's direct radiant energies graced the earth during the first three days of creation. • Through God's foreknowledge of the entire human existence on the earth, God made accommodations for the sun to take over from God's direct radiant energies on Day 4 of creation.
2nd	From Day 4 of Creation to The Original Sin	Perfection	• God created the universe, and replaced His direct light on the earth with the sun, the moon and the stars. • Weather conditions on the earth were perfect and allowed mankind to live comfortably in the open field.
3rd	From The Original Sin to The Great Flood of Noah	Sinful Nature Begins - Human Disobedience of God unleashed Activated Evil upon the whole world; and all natural genetics were altered to incorporate sinful nature. Rot, Decay and Death set in.	• Mankind went from perfect Creation to sinful nature. • Human Beings evolved. • Plants and animals independently evolved, also. • Conditions of the earth got progressively harder because there were no rains, and the heat of the sun became increasingly scorching.
4th	From The Great Flood of Noah to The Present Time	Sinful Nature Continues - God's renewed Grace out of God's Great Mercy, in the face of sinful nature.	• God sent First ever Rainfall on the Earth. That rainfall caused the flood that destroyed the old earth and replenished it with fresh fertile soil from the flood. Modern Mankind starts anew through Noah. • God created the rainbow. • Mankind began to communicate in different languages • Mankind was dispersed across the whole earth • The Earth was divided into Continents

Figure 9: The different Ages of the Earth

The 2nd age of the earth seamlessly continued from the first age of the earth. The only difference between the two ages was that while the first age was supported by God's direct presence on the earth, the second age was supported by the sun, the moon and the stars, with the universe lodged between the earth and God's dwelling place.

In both the 1st age of the earth and the 2nd age of the earth, living conditions in the open earth was perfect that human beings, and all the animals of the earth, did not have any need for shelters because there was nothing to seek shelter from. God had created perfect living conditions everywhere on the earth, and had maintained those conditions through the transition between the two ages.

Day 4 of creation was the start of the **2nd age of the earth** because on Day 4 of creation God replaced His direct presence on the earth with celestial elements, namely the sun, the moon, the stars and everything else that resulted from the cataclysmic explosion that we popularly call the Big Bang—*Genesis 1:14-19*.

The 2nd age of the earth seamlessly continued the conditions that existed on the earth at the end of Day 3 of creation. God had just unfolded the universe over the earth and blocked the earth's view of heaven completely. Hear it from the Bible:

"__Is not God in the heights of heaven? And see how lofty are the highest stars!__ [13] Yet you say, 'What does God know? __Does he judge through such darkness?__ [14] __Thick clouds veil him, so he does not see us as he goes about in the vaulted heavens.__" (Job 22:12-14)

At the end of Day 3 of creation anyone on the earth could see God with their naked eyes, seated majestically on His throne in heaven because of the infinite size of God. In contrast, by day's end on Day 4, that view of heaven was completely blocked by the universe that God put in place on Day 4.

The moon's reflection of the light of the sun was to point mankind to the reality that the sun and the stars in the universe continually receive their radiant energies directly from God and His Christ—and not from some imagined infinite supply of hydrogen gas within each star.

This is the Supremacy of God, and a demonstration of God's perfect wisdom. God can lead you to illuminating knowledge; or He could lead you to wherever else you convince yourself to go just to avoid God and instead elevate yourself.

The expansive universe being placed between the earth and heaven was no minor event. It was the beginning of a new age for the earth—a planet God had created to have eternal existence, just like heaven has eternal existence. The earth is truly in its own class, and will never be rivaled by any other celestial body.

World scientists have been scouring the entire universe, frantically searching for another earth to prove that the Bible is wrong. And all of us in the world are cheering them on in the name of the pursuit of knowledge—even though their practice is nothing but absurdity.

God was not mistaken, therefore, when He prophesied that all the people of the world will worship the beast. The world has got another thing coming. No one who sets out to discredit God will ever succeed. And believers who quietly cheer for these maddening escapades are walking on slippery slopes.

Before the universe came into existence, there was heaven and the earth *(Genesis 1:2)*. And when it is all over—at the end of time—there will, once again, be heaven and the earth. Whereas the very first earth had seas, the earth at the end of time will have no seas, but rather an unfettered access to the crystal water of the "New Jerusalem".

The second age of the age started from Day 4 of creation and continued through the days of Adam and Eve in the Garden of Eden, up to their rebellion against God. Everything in the world during the second age of the earth remained perfect, because there were no activated evil to mess anything up. Adam and Eve obeyed God's commands and satisfied God completely, and enjoyed perfect security in God's continued presence in the world.

Then came the rebellion—the rebellion of Adam and Eve against God. And the evil which God had made inert to mankind became instantly activated by mankind's disobedience, and was unleashed upon the entire world. Rot, decay and death set into everything that was in existence in the world.

The time period from the rebellion of Adam and Eve at the Garden of Eden to the Great Flood constituted the **3rd age of the earth**. The third age of the earth was an age of difficulty in the world. Grace was greatly diminished by sin, and activated evil took up any room vacated by Grace in the world. Death had set in and the world had begun to die. Mankind became more sinful and activated evil continued to accumulate in the world and cause havoc to lives throughout the world.

All natural genetics were reconstituted through the activities of the activated evil (voided Grace). This all became possible by the withdrawal of Grace from within each living thing to the degree predestined by God as a detriment for wickedness and evil. The withdrawal of Grace creates room for activated evil to move in to the extent which Grace was withdrawn from each living thing. And sinful nature became a norm in the world.

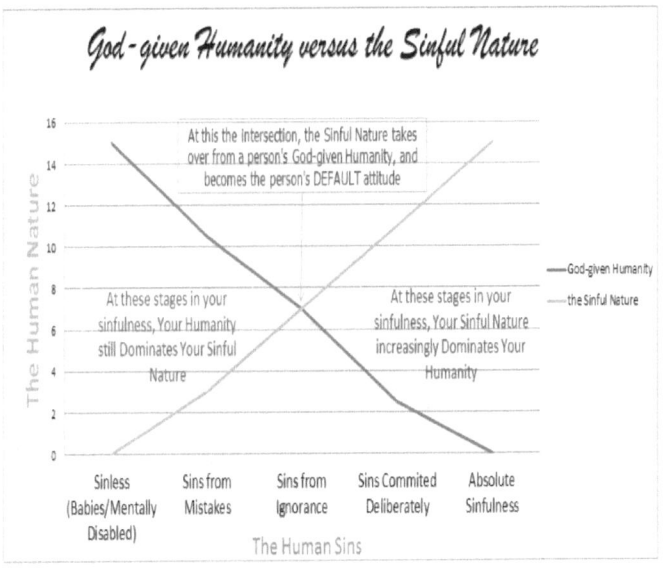

Figure 10: No human person ever came into the world with the sinful nature—not even Cain. God makes sure that every human baby that enters into the world, comes in with God's full love—which is a human person's perfect humanity.

The heat of the sun gained in intensity and began to scorch the earth and all of its content; bringing severe and irreversible changes to the world's weather, and creating adversities that freely impacted all the living things on the earth. It is possible that the oscillation of the earth—23 degrees north and 23 degrees south—started at this point in time, thereby creating different seasons of the year.

The animals and the plants of the world were all adversely affected, all undergoing genetic mutations that caused all livings things to begin to evolve according to God's predestinations. And Grace (God's Free Spirit), which is abundantly everywhere in the world and in the universe, monitored, controlled and generated the desired results.

With no rain on the earth since God created the earth, and the heat of the sun now scorching the earth and everything on the earth, life was becoming truly unbearable on the earth. With mankind driven out of the Garden of Eden and forced to farm to support itself, things really got progressively dire, prompting Lamech to prophecy that God may use his son, Noah, to bring relief to the world.

And God did bring relief to mankind through Noah, by covering over everything sin had tainted. New earth's surface and enriched soil, regular rain to water the entire earth's surface and replenish the ground water, new breeds of animals, new plant stock, and God's renewed Grace were given to Noah and his family to start over.

God put an end to the third age of the earth by covering the entire earth and everything on the earth—even the highest mountain peaks on the earth—with flood water that lasted almost a year over everything. The flood replaced the surface of the original earth with a new layer of earth throughout, greatly changing all of earth's geology.

Following the great flood—about 105 years after the flood—God divided the earth's landmass to create the continents that we see on the earth today, further affecting the earth's geology, climate patterns and wind patterns, among other things.

The **4th age of the earth** started when Noah's ark settled back on the surface of the earth and all the passengers got out and spread out across the new surface of the earth. That fourth age of the earth had continued till this day. And God promised never to destroy all life on the earth with water again.

Everything the world scientists use as their references comes from this fourth age of the earth. But the fourth age of the earth does not reflect the harsh conditions on the earth during the third age of the earth—that is the age between the original sin and the destruction of the earth by flood. Nor do the prevailing conditions in this fourth age of the age, in any way whatsoever, reflect the perfect conditions that existed on the earth during the first and second ages of the earth.

God did not bring any rains on the earth from the time of the earth's creation until the rain that destroyed the entire life on the whole earth. But the world scientists had made the assumption that everything we see in the world today has always been a part of the world since the earth came into existence. That is shortsightedness, especially when God presented the entire truth to us in His Holy Book—the Bible.

The world scientists have no way of theorizing or deciding what the first age, the second age, or even the third age of the earth, were like—their distinctions and their similarities. As such, no world scientist, or anybody else at that, is in a position to even remotely describe the conditions that existed on the earth on the three previous ages of the earth, aside from the information which God provided to us in the Bible, from which I put these things together.

Things from those three ages could only be understood through God's account of His creation activities in the Bible. And to those who do not believe God and have shunned His truths, those events would forever remain mysteries. The world scientists can make all the conjectures they want, but none of that would change anything in their favor. I guarantee that with every breath in my body!

There is a saying in my native African language which says that anyone who was not present at a burial; who decides to unearth the corpse; always starts at the foot position, instead of the head, because they usually could not tell which is which. That is dead-on with the wisdom of the world—confusion of the mind!

People often claim that they know something that they don't, because they have sold other lies to everyone before; and were rewarded handsomely for their deceits! To continue to reinforce their importance, they continue to manufacture fantasies that are believable. And those, whom the fantasies are meant for, will continue to eagerly consume them, because they are already hooked. It is common human tendency due to constant desire for worldly pleasure for human beings to try to avoid God, because human beings enjoy darkness more than the light—the light exposes their evil schemes *(John 3:20).*

God says that water and the earth were created before anything else in the entire universe. But scientists could not prove how that is possible with God removed from creation. So, they devised a roundabout way that puts the universe first. The Big Bang, by the way, originated in the Bible—deciphered by a brilliant Roman Catholic priest, George Lemaitre.

In their roundabout theories, the universe, through magical transformations and incantations, then produced the earth and all the lives that exist on the earth. They branded their science-fiction, a science; and continue to feed that to everyone in the world through government-sponsored programs and public school systems.

Because they control all of the world's resources, and all the media outlets, whatever they say anything is, is generally accepted as what the thing is—at least until they change their mind about it. But we are all losers by continuing to accept those senseless concoctions without checking them against the truths of the Bible for accuracy. Using these fallacious conclusions as the standards for truths is detrimental to all mankind—we are all worshipping the beast. Yet nobody seems to care.

Everything mankind has experienced in this fourth age of the earth is an experience already tainted by sin. Mankind, and all the plants and all the animals on the face of the earth, started the fourth age of the earth with a sinful nature. None was any longer the perfect creation which God created during the six days of creation in Genesis Chapter One.

The biological food chain started at the very beginning of the fourth age of the earth—Noah's family's return to the earth after the Great Flood. Animal killing, meat eating, animals preying on other animals, all started at the onset of this fourth age of the earth. Our current world is a world that has been reconfigured to now accommodate the sinful human beings and their untamable desires for selfish pleasures and comfort.

Chapter 2

God created the Earth and the Universe!
(Shout it from the Mountain Top! Shout it everywhere!)
(And teach it in every classroom in the world!)

Before Time existed, God created Water & the Earth. Then He created **Light**. Earth's **gases** started to form through **ionization** by this **Original Light** that God had dawned on the earth. **Time** started ticking—first the noon; then, the evening; and finally, the morning to complete **Day 1**.

On **Day 2**, God made **spherical space** appear inside the water which surrounded the earth, thereby separating the water underneath the space from the water atop the space. Then, He stretched out the spaces, creating under vacuum, the seemingly infinite expanse that is **space**. And the water atop this expansive space still exists today beyond space, according to the Genesis account.

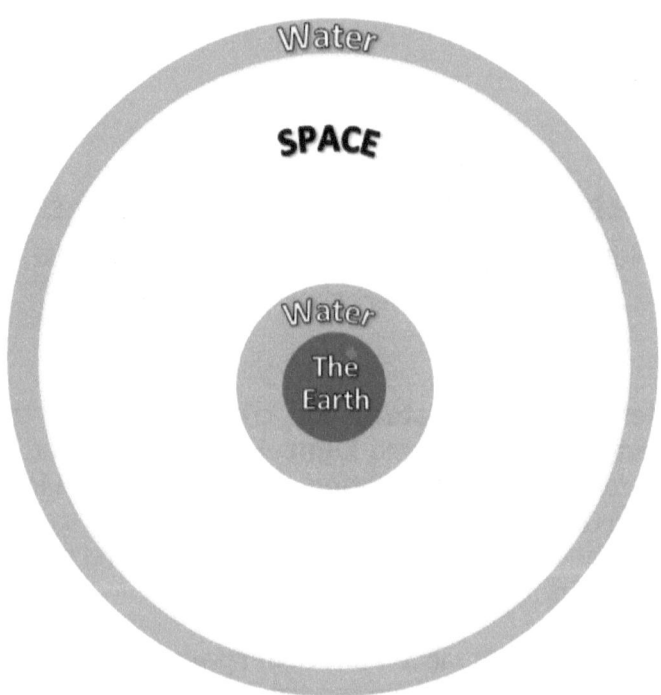

Figure 11: God created space on Day 2, pushing water to the outer limits of space to create space under unfathomable vacuum that helped dispersed the celestial bodies that resulted from the Big bang on Day 4 of creation.

On **Day 3**, God created **land** out of the water still surrounding the earth through an unimaginably huge **volcanic** explosion. **Tsunamis** ensued and cooled and weathered the land. Rivers, streams and lakes were formed. Earth's gases were completed and earth's atmosphere sealed off and protected by the Spirit of God. And God created vegetation throughout the land.

On **Day 4**, with just one command, God set off the

Big Bang through the Original Light of Day 1 through Day 3, and the universe was born, aided by the vacuum.

God's Equation: God is infinite energy, plus the sum total of all the energies that exist everywhere in the universe, and on the earth, visible and invisible; thermal, nuclear, sonic, light, magnetic, electrical, chemical, potential, hydrostatic, and all the other forms of energy known and unknown. And all these are summarized in the following elementary equation as:

$$E_{God} = E^{\infty} + \sum_{n=1}^{\infty} E_n$$

Where:

$E_{God} = \text{God's Energy};$

$E^{\infty} = \text{Infinite Energy};$

E_n=Total Energy of its kind in the universe

n=number of energy types available in the universe, known and unknown; visible and invisible.

And **Before Time** Existed *(Genesis 1:1-2):*

God created the Earth as a hot, molten mass *(Job 38:14)* submerged inside a deluge of Water *(Genesis 1:2).* Dense water vapor *(Job 38:9)* rose from the quenching of the hot mass (the molten earth) and caused thick darkness over the surface of the water which was covering the earth *(Genesis 1:2).* And Gravity (the Spirit of God) held everything together *(Genesis 1:2).* This darkness symbolized the darkness of the heart that would soon after creation descend on mankind and the world which can only be removed through God's grace. It must be inferred then that **matter** was created before time, space and the universe.

And as God was crafting His masterpiece, the angels watched and celebrated: Here is the Scripture: *"Where were you when I laid the earth's foundation? Tell me, if you understand. ⁵ Who marked off its dimensions? Surely you know! Who stretched a measuring line across it? ⁶ On what were its footings set, or who laid its cornerstone—⁷ while the morning stars sang together and all the angels shouted for joy? (Job 38:4-7).*

Day 1 of Creation *(Genesis 1:3-5):*

God brought forth the **light** and darkness receded. Light was not created on this day. Light was made to shine on the earth for the first time ever on this day.

And God was the source of that light— *"God is light; in him there is no darkness at all" (1 John 1:5);* which makes sense because the Origin of light Himself (God) and the heavenly hosts who were present at the creation of the earth and the universe would not have been in darkness before this time. Therefore God has had light for all eternity because that is His nature. And God promises to be mankind's direct light again at the end of this age when the current universe passes away and the earth is renewed, *Revelation 22:5.*

God shined on the earth from Day 1 to Day 3, through His Son Jesus Christ (John 1:1-5). And since God made the earth spherical, the light illuminated one half of the sphere that is the earth while the other half was in darkness. The earth started to rotate so that every portion of it would receive the light within a 24-hour period. The appearance of light marked **noontime** on that first day. The **evening** came and **morning** followed, thus completing a full **day**—the first day.

The earth's gases started to form through ionization by the True Light. The Bible says that God is a consuming fire *(Hebrews 12:29—"for our God is a consuming fire").* Therefore no celestial body has the ionizing power that equals God's—God is infinite energy plus the sum total of all the energies that exist everywhere in the universe, known and unknown, visible and invisible.

The dawn of light on the earth marked the beginning of **time** in the universe since the earth was created before time, and the universe was created on Day 4 of creation.

Day 2 of Creation *(Genesis 1:6-8):*

God created a void inside the water which totally encapsulated the earth and some water *(Genesis 1:6)*. This is a spherical void; and this void separated water underneath the void from water on top of the void. God "stretched out" the void to form the current-day space *(Genesis 1:7 & Isaiah 45:12—"**I have made the earth, and created man upon it: I, even my hands, <u>have stretched out the heavens</u>, and <u>all their host have I commanded</u>.")*. God called this space heaven(s) (Genesis 1:8). This space at this point was under vacuum since it was entirely created inside water. And the water on top of the "firmament" was pushed out of the universe; still surrounds the universe till this today.

Day 3 of Creation *(Genesis 1:9-13):*

At God's command *(Genesis 1:9)*, the following events took place: Land came out of the water through volcanic eruptions. Water channels got trapped inside the earth, cooling the earth and forming water reservoirs to feed rivers and streams *(Genesis 2:5-6)*.

Tsunamis ensued, violently clapping over the new land—cooling it, wetting it and weathering it. As land formed, water was pushed to one side to form one huge sea.

{When God destroyed the earth in the days of Noah, He had opened the earth and allowed some of the trapped water to come upon the earth's surface and create great flooding and devastated the earth. In Noah's flood, the earth's crust severally ruptured to let trapped water out to the surface of the earth; causing the land to break up into huge segments.}

{These segments later moved away from one another to form the various continents—the continental drift— *Genesis 10:25—"One was named Peleg, because in his time the earth was divided* ..." Peleg was the 5th generation from Noah—born 201 years after the flood.}

At God's command *(Genesis 1:11)*, the land that was formed, cooled and weathered earlier in the day produced vegetation *(Genesis 1:12)*: The land produced grass and herb yielding seeds, and the fruit trees, whose seed was in itself, all after their own kinds.

{We should all remember that God made a plant grow overnight and produced enough foliage to provide shade for Jonah to stay under while Jonah was protesting God's leniency with the people of Nineveh. The plant then died by dawn the next day, arousing greater anger in Jonah *(Jonah 4:10—"It sprang up overnight and died overnight.")*}

The earth's atmosphere was completed between Day 1 and Day 3, since on Day 3 God created plants on the earth which depends on minerals from watered earth; gases from the atmosphere; and light energy from the light—the True Light that dawn on Day 1 and lasted to Day 3.

The Garden of Eden was planted by God on Day 3 to accommodate and provide for the man and the woman God would make on Day 6 of creation. *(Genesis 2:8-14)*—God planted all vegetation on Day 3.

The formation and configuration of the earth, its seas, its atmospheres and everything contained within them is complete by the end of Day 3, and sealed off by the Spirit of God, in preparation for the Big Bang that follows on Day 4.

Day 4 of Creation *(Genesis 1:14-19):*

God's command for lights in the expanse of heaven *(Genesis 1:14-19)* set off the Bing Bang, and trillions of oceans of light and fire flew in all directions across the space God had created on Day 2 *(Genesis 1:6-8)*, filling it completely; aided by the vacuum God had created within space. {Space was created inside water—separating water from water—thereby drawing a vacuum; with the earth, its life, its seas and its atmosphere at the core of the space, held together and held in place by the Spirit of God (gravity)}.

The Spirit of God also protected the earth, its fragile vegetation and its atmosphere from being incinerated and completely consumed by the explosive energy of the Big Bang. This is God showing His infinite capacity to contain and protect whatever He wants to contain or protect.

Through the Big Bang God created the universe and all the celestial bodies. But the Big Bang did not create life, the light, the earth, earth's seas (water), and earth's atmosphere (gases). And up to this day, there exists water outside the perimeters of—and completely surrounding—the current universe, according to *Genesis 1:7*. The firmament God called the heaven—and we today call space—separated water from water.

The fires and lights are imprints of the True Light which set off the Big Bang; and were arranged in clusters; set in orbits; contained by gravities; and given <u>revolutionary</u>, <u>oscillatory</u> and <u>rotational</u> motions to serve God's various purposes.

The earth started to revolve around the sun to create the year. It also started to oscillate, 23 degrees to the south and 23 degrees to the north, to create seasons. And the universe as we know it today was born. God positioned the sun to take over from the True Light of

Day 1 through Day 3 of creation as the source of light on the earth.

He also created the moon to reflect the light from the sun and illuminate the night; and more importantly to serve as a sign pointing mankind to the sun and the stars being imprints of the True Light. God created the stars and completed an intricately balanced universe.

Day 5 of Creation *(Genesis 1:20-23):*

On Day 5 of creation, God created all sea animals and all the birds. He commanded for them and acted on His command *(Genesis 1:21-22).*

Day 6 of Creation *(Genesis 1:24-31):*

On Day 6 of creation, God created all land animals *(Genesis 1: 24-25).*

And finally, God created man. He created both man and woman on this Day 6 of creation: Adam was first and Eve followed later that day—*"So God created man in his own image, in the image of God created he him; <u>male and female created he them</u>. And God blessed them, and God said unto them, Be fruitful, and multiply, and replenish the earth, and subdue it: and have dominion over the fish of the sea, and over the fowl of the air, and* *over every living thing that moveth upon the earth. And God said, Behold, I have given you every herb bearing seed, which is upon the face of all the earth, and every tree, in the which is the fruit of a tree yielding seed; to you it shall be for meat. And to every beast of the earth, and to every fowl of the air, and to every thing that creepeth upon the earth, wherein there is life, I have given every green herb for meat: and it was so." (Genesis 1:27-30).*

"And God saw every thing that he had made, and, behold, it was very good. <u>And the evening and the morning were the sixth day.</u>" (Genesis 1:31)

"Thus the heavens and the earth were completed in all their vast array." (Genesis 2:1)

Day 7 of Creation *(Genesis 2:2-3):*

"By the seventh day God had finished the work he had been doing; so <u>on the seventh day he rested from all his work.</u> ³ Then God blessed the seventh day and made it holy, because on it he rested from all the work of creating that he had done." (Genesis 2:2-3).

GENERAL OBSERVATIONS:

I like to point out, at this point in our discussion, an interesting observation about the records of God's creation in the Bible: When God commanded for something which were to result from something He had previously built into what He was commanding it out of, what He commanded for simply happened as God had commanded; without God performing additional act to accomplish it; thereby suggesting a natural progression.

In these instances of natural progression, God's commands activated what He already built in, and what God commanded for automatically materialized as God wanted it, without further action by God. It appears that the code for creating what God commanded for was already imprinted on the source from which the result came forth.

For example, when God commanded for there to be light in *(Genesis 1:3)*, light automatically appeared. God did not act for the light to appear as He had commanded. The light simply came forth from its source—a source which God had previously made. God's command simply activated the source and light shined.

That is also the case with the formation of land out of the water and the separation of the sea from the land where through the progression of natural events, God's commands were accomplished *(Genesis 1:9-10)*:

At God's simple command, massive volcanos jetting through the huge standing water and created new expansive land mass, displacing equal volume of water, which naturally raced away from the disturbance only to return violently and overrun the new land mass; and flowed away once again, and returned yet again; cooling the land, watering it and weathering it.

And when everything settled down, the new land stood majestically over the water, with huge reservoirs of water trapped within the land, which originated rivers, lakes and streams and continued to water the land. The rest of the water naturally collected around the land, forming the sea.

Science has concluded that 98% of the earth's land is of volcanic origin. It is not 98%. It is all of the earth's land that was created through volcanos by God on Day 3 of creation in *Genesis 1:9-10.*

 This natural progression also applied in the case of vegetation coming forth from the land. In this instance, it appears that the genetic codes for the various plants that grew out of the land were already part of the constitution of the land, waiting to be activated by the voice of God. And naturally, when vegetation dies, it goes back into the earth.

Only God's command made the vegetation come forth. God did not have to follow His command up with any other actions.

And once the vegetation came forth, it was supported by the natural elements which God had previously created to support vegetation—**water** (created before time existed), **light** (created on Day 1), **gases** (created between Day 1 & Day 3), **land** and **nutrients** (created on Day 3).

On the other hand, when God commanded for something and had to create the thing from the scratch, the Bible records that God followed His command up with some action; and then the intended result happened.

This applies to the creation of the firmament (space) *(Genesis 6-8)*. We learn from *Isaiah 45:12* that God actually stretched out the space under vacuum after He commanded for it to appear, and because of that the Bible says: *"So God made the vault and separated the water under the vault from the water above it. And it was so." (Genesis 1:7).*

This is a project of unimaginable proportions and unfathomable science. God commanded for it to happen. Then He went on and made it happen. And in the process, God pushed the water on top of the space out of this universe, where it still exists, since the Bible account did not say that this water ever rejoined the water it was separated from.

The Bible tells us that God separated water from water in this process. And naturally, if and when anything is stretched entirely out inside water, it pulls a vacuum within its inside space.

For instance, stretching out a sealed and deflated balloon inside water would cause the balloon to elongate beyond its original size, significantly increasing the inside surface area of the balloon.

If we were somehow able to take away the elasticity of this balloon at the current extended state, and transform its walls to rigid walls, and pull the walls away from the center of the balloon, we would now have a rigid object that is larger than what we started with which has a vacuum inside of it.

And if we had let some water into the balloon and tied its end before stretching the balloon, and did everything else we described in the preceding paragraph; and if we had a method of keeping the water inside the balloon suspended in the middle of the space within the balloon, we would end up keeping the water inside the balloon well away from the water outside the balloon.

This is exactly the scenario the Genesis account of creation is talking about. In creating space on Day 2, to accommodate the Big Bang and the resultant universe on Day 4, God separated the water originally surrounding the molten earth at Genesis 1:2, into two parts.

One part of the separated water remained around the earth which still sat inside the newly created space, at the center of it. And the second part of the separated water, that which surrounded the outside of the space, was completely pushed out of space and never re-entered space, thereby leaving space in a state of vacuum.

This vacuum aided in the instantaneous dispersion of the celestial bodies that God shaped and organized immediately following the Big Bang.

As these bodies raced away in every direction from around the earth, God arranged them in clusters to form galaxies; set the galaxies and the individual bodies within each galaxy in orbits and inter-relationships; and set their motions and functions.

So, the creation of the universe through the Big Bang in *Genesis 1:14-19* (Day 4 of creation) started first by God's command which set off the explosion. Then God again followed His command up with action.

And for that, the Bible says: *"And God made two great lights; the greater light to rule the day, and the lesser light to rule the night: he made the stars also. And God set them in the firmament of the heaven to give light upon the earth, and to rule over the day and over the night, and to divide the light from the darkness: and God saw that it was good." (Genesis 1:16-18).*

While God was making and dealing with these extreme energies, He protected the fragile earth and its infrastructure from being vaporized by the excessive energies generated in the Big Bang; thereby exhibiting the limitlessness of His capacity.

On Day 5 and Day 6 of creation, God also followed His commands with action:

For the sea animals and the birds, the Bible says: *"And <u>God created</u> great whales, and every living creature that moveth, which the waters brought forth abundantly, after their kind, and every winged fowl after his kind: and God saw that it was good. And God blessed them, saying, Be fruitful, and multiply, and fill the waters in the seas, and let fowl multiply in the earth."* (Genesis 1:21-22).

And for the land animals on Day 6, the Bible says: *"And <u>God made</u> the beast of the earth after his kind, and cattle after their kind, and every thing that creepeth upon the earth after his kind: and God saw that it was good."* (Genesis 1:25).

God created the animals and gave them breath and mobility. The animals and man were formed from the wet earth, once again suggesting that the earth already contains the constituent genetic codes to form the animal bodies: the cells, the tissues, the organs, the bones and the blood.

Biology speaks about dormant viral genes that pervade the earth's landscape which once inside a host's living cell gets activated and takes over the cell's anabolism and decimates the cell.

Dormant things do not all of a sudden become alive. God activates them like He did in the very beginning. These dormant biological pieces remain part of the earth and only become active by the Spirit of God who gives life to everything.

So in God's creation of animals, these constituent genetic codes were not only waiting for God's commands to be activated, they needed God's hands to form them; once again, signifying these creations to be something entirely new—just like **space** and the **universe**, which God formed and positioned with His hands.

In *Genesis 2:19,* the Bible repeated that God did the forming: *"Now <u>the LORD God had formed out of the ground all the wild animals and all the birds in the sky.</u> He brought them to the man to see what he would name them; and whatever the man called each living creature, that was its name."*

And when it finally came time to make man, God did not just form man like He formed the sea animals and the land animals.

He decided to form man in the image and likeness of God, and gave man His Spirit so that He could keep man in perpetual communion with Him. God did not simply speak and man came forth.

He formed man with special considerations. That is why the Bible says that man was fearfully and wonderfully crafted, indicating that man was designed to dominate his world just like we do today. Everything we do is not by chance. It is the design of God and our assigned destiny.

In conclusion, those things that God created, whose constituent genetic codes were already inherent in their source, simply got activated by God's commands and materialized as God commanded for them to materialize.

And those other things that God created, whose constituent genetic codes were already in their sources but are coming into existence for the very first time, or require God's special consideration, God's commands for them were usually followed up with action from God for them to materialize as God commanded for them to.

And in the case of human beings, God said, *"Let us make man ..."* and not *"Let there be ..."*, indicating that humans were made with the uttermost consideration of all by the God of the universe. What an honor!

NOTE: *Genesis 1:1 – "In the beginning God created the heavens and the earth."*— is a thesis statement about God's entire creation activities of the earth and the universe. The verse is not an itemization statement about the heavens being created before *Genesis 1:2* because the heavens were not created until Day 2 when God created space.

And on Day 4, God populated space and the universe was born. In much the same way, *Genesis 2:1—"Thus the heavens and the earth were completed in all their vast array."*—is a summary statement about the same events.

In essence, The Earth and water were created in

advance of Day 1. On Day 1, God shined His light—the True Light—on the earth, and this light lit the earth through Day 3.

Then on Day 4, the sun, the moon, the stars and the rest of the universe were born through the so-called Big Bang, with the Spirit of God controlling the whole event and protecting the earth at the same time.

Figure 12; The earth submersed inside the water of creation.

Figure 13: Appearance of light on the earth ushered in time.

Day 1 - God Created Time by Dawning Light

Genesis 1:3-5

At God's command, light dawned and earth's gases were formed through ionization by the light.

The Earth started to rotate so that the light would touch every part of the earth within each 24-hour period. In essence, the earth makes one full rotation every 24 hours.

The earth's atmosphere was completed sometime between Day 1 & Day 3, sealed off and held together by the Spirit of God – gravity.

Genesis 1:3-5. And God said, "Let there be light," and there was light. [4] God saw that the light was good, and he separated the light from the darkness. [5] God called the light "day," and the darkness he called "night." And there was evening, and there was morning—the first day.

Figure 14: Day 1 of creation.

Figure 15: With water on top of space, and water underneath space, space was under unfathomable vacuum when it was finished by God on Day 2 of creation.

Day 2 - God created Space

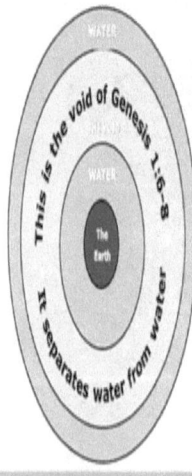

Figure 2: (**Genesis 1:6-8**): And God said, Let there be a firmament in the midst of the waters, and let it divide the waters from the waters.

[7] And God made the firmament, and divided the waters which were under the firmament from the waters which were above the firmament: and it was so.

[8] And God called the firmament Heaven. And the evening and the morning were the second

Genesis 1:6-8

God introduced a spherical void inside the water, separating water from water, with the earth and its seas trapped under the spherical void, and the rest of the water atop the spherical void. By implication, there is much more water outside the universe than there is on the earth.

The result of the introduction of this spherical void was four concentric spheres:

- the earth is the innermost sphere,
- water (which later became the seas) wraps around the earth,
- the void wraps around the water and the earth, and
- more water wraps around the void, the water and the earth.

Figure 16: Day 2 of creation.

Figure 17: This is a photo of a real island that was forming from an underwater volcano in the southern Pacific Ocean—New Tsonga Island.

Day 3 - God Created Land out of the water, pushing the water to one side

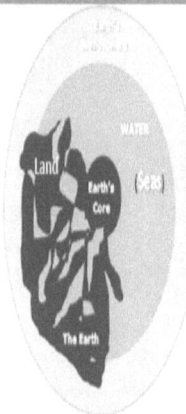

Genesis 1:9-10. And God said, Let the waters under the heaven be gathered together unto one place, and let the dry land appear: and it was so.

[10] And God called the dry land Earth; and the gathering together of the waters called he Seas: and God saw that it

Genesis 1:9-10

Land came out of the water through volcanic eruptions. Water channels got trapped inside the earth, cooling the earth and forming water reservoirs to later feed rivers and streams. Tsunamis ensued, violently clapping over the new land—cooling it.

When God destroyed the earth in the days of Noah, He opened the earth and allowed some of the trapped water to come upon the earth's surface and flood it.

In Noah's flood, the earth's crust severally ruptured to let trapped water out to the surface of the earth, causing the land to break up into huge segments.

The segments later moved away from away from one another to form the various continents—the continental drift—*Genesis 10.25—"One was named Peleg, because in his time the earth was divided ..."* Peleg was the 5th generation from Noah—born 201 years after the flood.

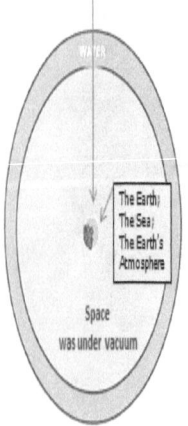

Figure 18: On Day 3 of creation, God created land out of the water of creation in the hollow of His mighty hand: *"The earth is the LORD's, and everything in it, the world, and all who live in it; [2] for he founded it on the seas and established it on the waters." (Psalm 24:2)*

Figure 19: God's command for light in the expanse of space sparked the Big Band, and the universe formed in space—which God had created over the earth and the earth's seas on Day 2 of creation. It took 24 hours to form the universe in space; from the beginning of it to the end of it *(Genesis 1:14-19).* **God is not a man that He should follow our thinking and reasoning. He is all-powerful and does not need up to 24 hours to form the universe. The universe took Him 24 hours because He had already allocated 24 hours for it. God does as He wishes, and never tries to prove Himself to anyone!**

Day 4 - God created the universe

On Day 4 of creation (Genesis 1:14-19), God's command for lights in the expanse of heaven set off the Bing Bang and trillions of oceans of light and fire flew in all directions across the space God created on Day 2, filling it completely, aided by the vacuum within the space.

The fires and lights are imprints of the True Light that set off the Big Bang; and were arranged in clusters, set in orbits, contained by gravities and given revolutionary, oscillatory and rotational motions to serve God's various purposes. And the universe as we know it today was born.

Genesis 1:14-19. And God said, Let there be lights in the firmament of the heaven to divide the day from the night; and let them be for signs, and for seasons, and for days, and years ... And the evening and the morning were the fourth day.

Figure 20: Day 4 of creation.

Figure 21: Here is a sample of our LifeCard, which is designed for your use in encouraging your loved ones with the wisdom of God.

ABOUT THE AUTHOR

My life is a laboratory. And all human beings are designed as such by the all-knowing God. The only difference among us is that while some willingly become part of life's experiments, some view it from the sidelines.

The best lessons we each learn in life comes to us directly and not through a teacher in an academic setting. We all learn and mature in our experiences by trial and error, just like a scientist in the laboratory. But we are not only the scientist, we are also the test specimen and the laboratory facility & instrumentation—all rolled into one.

And when we are in tune with our spirits, it becomes more verification than 'trial and error' because through our spirits, God feeds us great knowledge about our lives, the things around us and deeper mysteries than we ever thought possible.

Most of my books happened that way. Information came into my mind and takes residence. I soon become aware of it and try to know more about it. As I explore it, it deepens and more is downloaded onto my spirit. And intuitively, I am led to its verification. Once verified, it becomes common knowledge to me.

God has been unbelievably good to me by opening windows to me into great mysterious, such as I have been writing about in my many books. There is hardly a day that I am not writing books. I work on several titles simultaneously, capturing the information as soon as it enters my mind.

Ifeanyi Chukwujama

Other Titles form this Author:

- Who is God!
- What is Love!
- Christ is in Everyone!
- Christianity is not a Religion!
- The Singleness of God!
- Overcoming Your Trials!
- Live the Abundant Life!
- Science, Evolution and God!
- Reflections of Life!
- The Rapture, the Tribulations and the Church!
- The Big Bang: and Jesus Christ birthed the Universe!
- Government is a spirit and the Beast; Science is the False Prophet!
- Marriage is a Trinity with God!

www.ingramcontent.com/pod-product-compliance
Lightning Source LLC
Chambersburg PA
CBHW021411170526
45164CB00002B/606